P9-AOH-334

WHO AM I?

Who am I?

Titles in the series

Graceful and Galloping (Horse)
Heavy and Hoofed (Cow)
Loud and Crowing (Rooster)
Pink and Curly-tailed (Pig)

I am loud and crowing, proud and strutting.
I have a sharp beak.

WHO AM I?

By Moira Butterfield
Illustrated by Wayne Ford

Thameside Press

Distributed in the United States by
Smart Apple Media
1980 Lookout Drive
North Mankato, MN 56003

Printed in Hong Kong

Library of Congress Cataloging-in-Publication Data
Butterfield, Moira, 1961-
Rooster / by Moira Butterfield.
p. cm. — (Who am I?)
Summary: Describes parts of a well-known animal and invites the reader to identify it.
ISBN 1-929298-92-7
1. Rhode Island reds—Juvenile literature. 2. Roosters—Juvenile literature. [1. Roosters.] I. Title.

SF489.R6 B88 2000
636.5'84—dc21 00-022752

9 8 7 6 5 4 3 2 1

Editor: Stephanie Turnbull
Designer: Helen James
Illustrator: Wayne Ford / Wildlife Art Agency
Consultant: Jock Boyd

I strut around.
I’m stiff and proud.
My wake-up call is very loud.
I have long feathers in my tail,
So you can tell that I’m a male.
Who am I?

Here is my eye.

I strut proudly round the farmyard, looking after my family and watching for danger.

There are lots of animals living here. Can you find a cat, a mouse and two tiny ladybugs?

Here is my foot.

It has sharp claws
and a spike called
a spur at the back.
I slash enemies with
my claws in fights.

I am the only male
of my kind on the
farm. I fight any
other males like
me that come along.

Here is my beak.

I use it for pecking
at my food. The
farmer gives me grain
but I also look out
for seeds and insects.

I nibble on grass,
too. Sometimes
the farmer hangs
up vegetables for
me to peck at.

Here are my feathers.

You can tell that
I am male because
my tail feathers
are very long
and beautiful.

You can also tell
I am male because
I have a piece of skin
called a wattle hanging
down from my neck.

Here is my wing.

The farmer doesn't want me to fly away so he clips one of my wings.

When I am angry with another animal I puff up my feathers, flap my wings and rush forward to scare them away.

Here is my comb.

It is a brightly colored piece of skin sticking up from my head. Females have combs, too, but mine is bigger.

I live with a group of females. How many of us can you find in the farmyard today? Count the combs.

Here is my head.

When I want to signal to my family,
I open my beak wide and crow...

cock a doodle doo!

Have you guessed who I am?

I am a rooster.

I am called a
Rhode Island Red.

Point to my…

clawed feet

long tail feathers

large comb

red wattle

sharp beak

A rooster is a male chicken. A female chicken is called a hen.

Here are some baby chickens.

They hatch from eggs laid by the hens living on the farm. I don't lay any eggs.

Babies are called chicks. They follow their mothers around, making a tiny "cheep-cheep" sound.

Here is my home.

I live on a farm.

Look for me in the picture.
How many hens can you see?

Here are some other kinds of roosters.

◄ This rooster and hen are called Barnvelders. They are from Germany.

► These speckled chickens are called Marans. They live all over Europe.

This is a Japanese Yokohama rooster.
Look how long its tail feathers are!

Can you answer these questions about me?

Where do I live?

What color am I?

What do I like to eat?

Where are my longest feathers?

What do I do when I am angry?

What do I have
on my head?

What are baby
chickens called?

What kind of sound do I make?

Why can’t I fly away?

Can you name
some kinds
of roosters?

Here are some words to learn about me.

beak My mouth. I use my beak to peck at grain and seeds.

clip To cut or trim.

comb A piece of skin that sticks up from my head.

crow The sound I make. Can you make a crowing sound, too?

grain Seeds from plants such as wheat.

hatch To break out of an egg.
Baby chickens are born this way.

hen A female chicken.

spur A spike on the back of my leg.

strut The proud way I walk, showing
everyone how important I am.

wattle A piece of skin hanging
down from my neck.